1945

Creativity and Crisis: Chicago Architecture and Design of the World War II Era

The Art Institute of Chicago

1945: Creativity and Crisis, Chicago Architecture and Design of the World War II Era has been published in conjunction with an exhibition of the same title organized by the Art Institute of Chicago and presented from May 7, 2005, to January 8, 2006.

This exhibition is funded by the Fellows of the Department of Architecture with ongoing support from the Architecture and Design Society. Educational programs are supported by the Albert Pick, Jr., Fund.

First edition
Printed in the United States of America

Published by the Art Institute of Chicago
111 South Michigan Avenue
Chicago, Illinois, 60603–6110

This is a publication of the Ernest R. Graham Study Center for Architectural Drawing at the Art Institute of Chicago.

Library of Congress Control Number: 2005923393

ISBN 0-86559-218-7

Produced by the Publications Department of the Art Institute of Chicago, Susan F. Rossen, Executive Director
Edited by Robert V. Sharp and Elizabeth Stepina
Production by Sarah E. Guernsey
Designed and typeset by Jeffrey Wonderland, Department of Graphic Design, and Photographic and Communications Services
Separations by Professional Graphics, Rockford, Illinois

Except for the following, all photographs are by the Department of Graphic Design, and Photographic and Communications Services: p. 4: National Archives, Washington D.C., photo no. 80-GK-2862; p. 5: Top image from Bundesarchiv, Koblenz, Germany, 101/291/1213/12; bottom image from Boeing Historical Archives; p. 6: Seneca Public Library; p. 9: Top image from Chicago Bridge and Iron Company; bottom image from Henry Dreyfuss Associates; p. 27: National Air and Space Museum, Aeronautics Division Curatorial Files, Smithsonian Institution, Washington D.C.

Front cover, top: Aviation Machinists' Mates working on airplane engine, Navy Pier, July 1942. Photo: Naval Historical Foundation, Navy Pier. Front cover, bottom: James Edwin Quinn, detail of aerial perspective from Congress Parkway to Oak Park Avenue, 1945 (see p. 8). Inside front flap, top left: Albert Kahn, interior view of the Amertorp Factory, Forest Park, Illinois, 1943. Photo: Albert Kahn Associates. Inside front flap, center: Henry Glass, utility chair from the "Victory" series for W. L. Stensgaard and Associates, Chicago, 1943. Photo: The Art Institute of Chicago. Inside front flap, top right: Bruce Goff, interior view of the Ruth Ford House, 404 South Edgelawn, Aurora, Illinois, 1947–49. Photo: The Art Institute of Chicago. Inside front flap, bottom right: Daniel Brenner, detail of graduate student project at Illinois Institute of Technology for Ludwig Mies van der Rohe, 1946 (see p. 20). Inside front flap, bottom left: Bertrand Goldberg, exterior view of the Standard Homes, Suitland, Maryland, 1941–42. Photo: The Art Institute of Chicago. Inside back flap, top: James Edwin Quinn, detail of presentation perspective drawing of the Caldwell Bridge over the Edens Expressway, 1950. Hand-tinted photographic print on paper; 8 x 22.9 cm. Gift of J. Edwin Quinn, RX17563. Inside back flap, center: Model LST launching at Marshall Fields' *Victory Center*, July 17–August 20, 1944. Photo: Seneca Public Library. Inside back flap, bottom right: Chicago Bridge and Iron Company, waterspheroid tank in Elmhurst, Illinois, 1954 (see p. 9). Inside back flap, bottom left: The Austin Company, Consolidated-Vultee Factory, Fort Worth, Texas, 1942–43. Photo: The Austin Company. Back cover, top right: L. Morgan Yost, detail of aerial perspective of the Zoned House, 1944 (see p. 31). Back cover, bottom right: Richard Ten Eyck, Vornado fan, 1945–47 (see p. 28). Back cover, left: Grumman F6F Hellcat production, Long Island, New York, 1944. Photo: Cradle of Aviation Museum.

1945: Creativity on the Cusp between War and Peace

JOHN ZUKOWSKY

Certainly no one would dispute that 1945 was an important year in modern history, marking the end of a conflict that united many western democracies and some communist adversaries in a struggle for survival against the Axis powers led by Germany, Italy, and Japan. The rivalries that developed after World War II between the western democracies led by the United States and the eastern bloc nations led by the Soviet Union eventually led to a Cold War that lasted nearly a half century until the 1989 fall of the Berlin Wall.

Although the sociopolitical impact of these war years has been the subject of numerous books, lectures, films, and television shows, the contribution to the war effort by visual arts professionals—artists, architects, and industrial designers—has been barely touched upon. Perhaps it is difficult for us to imagine creative individuals involved in work that supports the destructive side of human nature. Perhaps these visual arts professionals have wanted to leave their horrific memories behind. As the history of any conflict is often written by the victors, it is comparatively easy to find out information about the wartime architectural works of prominent personalities such as Albert Speer, Hitler's architect and later armaments minister, and even less prominent architects such as Karl Bischoff and Fritz Ertl, who designed extermination camps at Auschwitz-Birkenau and elsewhere. But it is considerably more difficult to find any book on Erich Mendelsohn or Konrad Wachsman that talks about their role in the building of the U.S. Army's facilities in Dougway, Utah, where incendiary bombs were tested before they were unleashed on German and Japanese cities.[1]

The purpose of this publication and the exhibition it accompanies is to redress some of this neglect and examine selected instances in which architects and designers put their creativity to work in the war effort. In the following essays, we will establish that architects and designers—particularly those Chicago practitioners featured here—did indeed contribute to various national war causes and that what they did then influenced their immediate postwar work. We hope that our recognition of them will inspire historians to analyze similar work in other cities and countries, particularly in former wartime totalitarian environments such as Japan and the former Soviet Union, where access to information is often more limited than in the West. For now, it might help us understand the creativity of Chicago-based designers if we look at the wider context of what artists, architects, designers, and civil engineers did during the war years. Beyond that larger context, another aim of our project is to connect the bigger picture of these cataclysmic years to individuals whose stories remind us that real people participated in this heroic struggle on all levels.

Fig. 1 USS LCI–1097 and LST–652 open for visitors during the 6th War Bond Exhibit, Chicago, Illinois, December 1944. The LST–652, constructed from July to October 1944, was one of hundreds built at the Chicago Bridge and Iron Company shipyard in Seneca, Illinois (see fig. 4).

Architecture and Design for Modern War: Issues of Historiography

Some might rightly say that architects have always made designs for military purposes, among the earliest documented instances being the Roman city walls, catapults, and siege engines described in the *Ten Books of Architecture* by Vitruvius, an architect who built such military structures for the emperor Augustus. Similar structures and machines appear in the sketchbooks of the thirteenth-century French architect Villard de Honnecourt. And one need only be reminded of the spectacular medieval fortresses of the thirteenth and fourteenth centuries that ranged across what was then the known world, from the conquest castles on the coast of northern Wales to the walled city of Carcassonne in southern France to the Krak des Chevaliers, the dramatic Crusader castle in the Syrian desert. Countless other examples from the Renaissance and Baroque eras have been published, notably Leonardo da Vinci's sketches for military machines from the fifteenth century, Sebastiano Serlio's centrally planned polygonal fortress designs of the sixteenth century, and Sébastien Le Prestre de Vauban's star-shaped fortresses of the seventeenth century. Although there are numerous books that document this age-old architectural work for the military, it is difficult to find any monographs that survey similar architectural and design contributions in the twentieth century.

One of the few books to focus on the modern era is *The Architecture of War,* written in 1973 by Keith Mallory and Arvid Ottar.[2] Their survey examines the fixed fortresses of the interwar years, such as the French Maginot Line of the mid-to-late 1930s, which was intended to be the definitive border between France and her longtime rival Germany. The book also examines comparable reinforced concrete structures designed and created during World War II, including submarine pens along the French coast, antiaircraft bunkers in Germany, and the Atlantic Wall constructed by Germany's enormous design-build agency, Organization Todt, to repel any Allied invasion of France (fig. 2). The book also includes a number of creative civil engineering structures that were intended to be temporary, movable architectural solutions that could quickly solve specific logistical problems in a variety of circumstances. These include the famous Mulberry artificial harbors built off the French coast after D-Day, June 6, 1944, and the ubiquitous British Nissen huts (similar to the Quonset huts of their American counterparts) constructed to house Commonwealth troops all over the world as Allied armies mobilized to combat the Axis powers in the European and Pacific theaters. A particularly distinct example of creative design for these temporary structures is work by Bruce Goff, who, as an architect working with the Navy's Construction Battalion, or "Seabees," incorporated Quonset huts within some of his structures during and after the war (see pp. 14–17).

One could also think of the famous portable Bailey Bridges of 1944, named after British inventor Donald Bailey, who was knighted in recognition of the importance of his constructions to winning the war, and the large concrete castle keeps in Berlin, Hamburg, and Vienna from 1942 to 1944 designed by Friedrich Tamms that sheltered thousands of citizens during Allied air raids and housed antiaircraft guns, searchlights, and radar. As with many other concrete structures in occupied Europe, from submarine pens to underground factories, these massive constructions frequently proved too costly to demolish after the war, and they have been, at times, recycled into bars, film studios, offices, and galleries. Often the books that include such structures are targeted at enthusiasts or military history buffs and thus seldom identify the architects, designers, and

Fig. 2 Soldier in camouflage at the Atlantic Wall, 1943.

Fig. 3 John Graham. Aerial view of Boeing Plant 2 in Seattle during World War II, camouflaged with a canvas and wood suburban subdivision.

engineers who worked behind the scenes to create the infrastructure of war.[3]

Other important but underpublished architecture- and design-related subjects of the war years include military insignia, uniforms, and camouflage, the latter a fascinating subject of architectural constructions intended to deceive aerial photographic reconnaissance, even involving the construction of massive wooden decoy factories in Germany and occupied Europe.[4] As it happens, the Museum of Modern Art in New York organized an exhibition in 1943 on architectural camouflage and the School of the Art Institute of Chicago even taught courses in it, but only occasionally will one learn that a specific architect, designer, or artist was responsible for a camouflaged ensemble (fig. 3). Until scholars tackle all aspects of the subject, this will be a major missing chapter in the architectural and design history of the twentieth century. Only then will we be able to put into a wider context the better-known wartime work of, say, German factory architects such as Bernhard Hermkes and Herbert Rimpl and their American counterparts, Albert Kahn and the Austin Company, or even less well known industrial facilities such as the Seneca Yards of the Chicago Bridge and Iron Company (fig. 4). This engineering company, now based in Oak Brook, Illinois, constructed the Michigan Avenue Bridge (1918–20) and specialized in building water towers before and after the war (see fig. 8). But from 1942 to 1945 their shipyard on the Illinois River built hundreds of LSTs (Landing Ship Tanks) for the invasions of Europe and the Pacific (fig. 1).[5]

Fig. 4 Aerial photograph of the Seneca Shipyard where Chicago Bridge and Iron Company constructed LSTs during World War II.

Creativity in a Crisis: The Human Touch

When I first began organizing meetings for this project in spring 2002, I was struck by some of the observations made by representatives of veterans' groups who lived through the war as young adults. These individuals stressed the emotional aspect of the era—that it was indeed a struggle between life and death, between democracy and totalitarianism—and that 1945 represented the cusp between wartime fears and postwar hopes. Their statements reminded me that the project was linked to the experiences of real people, not bound by our cursory impressions of an era seen through black-and-white newsreel footage. Indeed, it is the stories that people shared in the making of this project that helped it come alive for all of us involved, and it was my hope that we could somehow incorporate this educational experience, in a very democratic way, into the exhibition.

As part of the planning process for the show, we decided to incorporate such first-hand experiences of "witnesses" of the era within Stanley Tigerman's installation so that those who lived through the era or were impacted by it could share their stories with the general public within the gallery. In some ways this might be considered a living extension of an oral history project such as the Art Institute's own pioneering program on Chicago architects, which includes reminiscences by several architects about the war years. Among these are a number that demonstrate creativity in the midst of crisis: Bertrand Goldberg, for example, discussed his work on temporary housing developed from field gun crates (fig. 5) and the influence his wartime housing work had on his postwar attitudes to system building (see pp. 10–13). A more incredible story is German architect Werner Buch's recollections of studying with Ludwig Mies van der Rohe in Chicago from 1939 to 1941, and Buch's return to Germany, months before America's entry into the war, to work on the design of modernist aircraft factories with Austrian architect Lois Welzenbacher. Buch's story is a truly amazing one; while he was a prisoner-of-war he was "rescued" by an American army officer whom he had known in Chicago, who then hired him to help begin the German radio news service. These are but two tales of many from the war that involve architects of the day, and the case studies in creativity that follow within this booklet and in the exhibition bring out the human touch, the human stories, in the design work of their creators.

1945: Creativity after the Crisis

The war's end brought with it the need for housing for returning troops and a pent-up demand for consumer goods. All of this was fueled by the lack of new homes and products dating back to the Great Depression of the 1930s, the fact that people had savings from their war earnings to spend, and the availability of new materials developed during the war. One need only recall feature films such as *Miracle on 34th Street* (1947) and *Mr. Blandings Builds His Dream House* (1948), both of which use the desire to escape the density of city living for rural or suburban single-family housing as part of their plotline. Events such as the *Chicago Tribune*'s Chicagoland Homes Competition of 1945 drew wide recognition and record-breaking crowds: over 90,000 people visited the exhibition at the Art Institute of Chicago in a single month in spring 1946, and when some of the winning designs were constructed, the model homes hosted over 212,000 visitors in one month. Similar housing exhibitions after the war could be found from London to Sydney, all showing the design dreams of living in a postwar

Fig. 5 Bertrand Goldberg. Cutaway perspective presentation drawing of a shipping crate for a Bofors 90mm antiaircraft gun that converts to housing for troops, c. 1943. Mixed media collage on board; 23 x 30.5 cm. From the archive of Bertrand Goldberg, gifted by his children through his estate, RX23664/161.1.

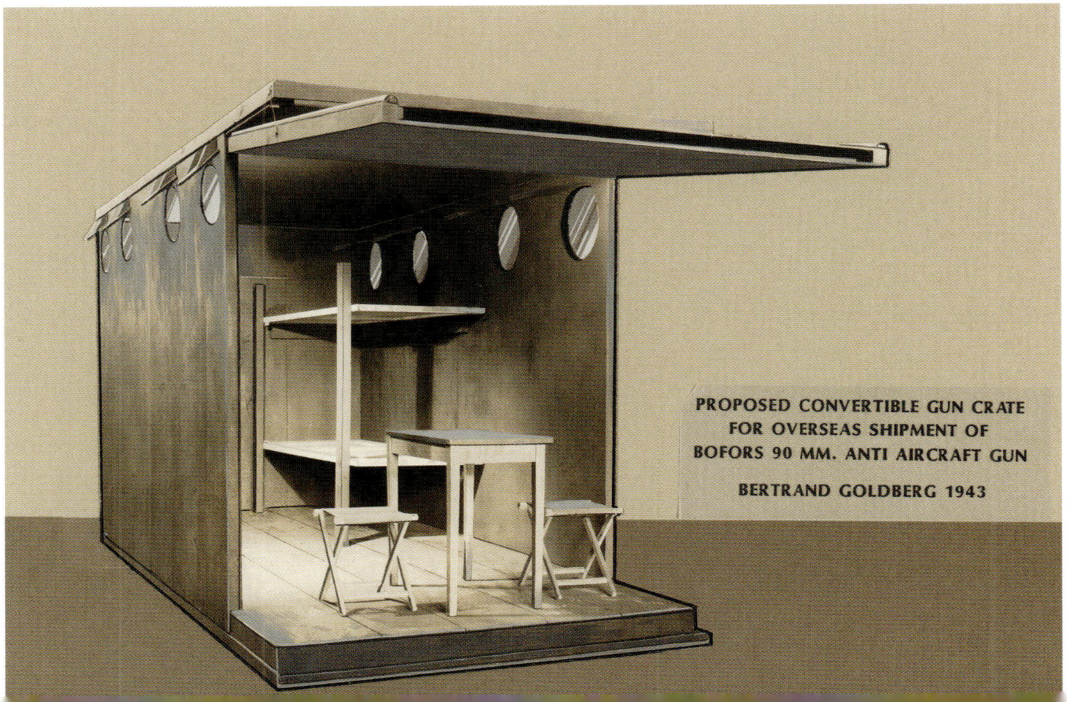

Fig. 7 James Edwin Quinn. Aerial perspective from Congress Parkway to Oak Park Avenue, 1945. Delineated by Al Burnes. Graphite and printed paper mounted on board, with ink, gouache, and watercolor; approx. 50.7 x 86.3 cm. Gift of J. Edwin Quinn, 1980, 1992.83.1.

Fig. 6 Advertisement for Swingline children's furniture designed by Henry P. Glass, 1952.

society. The development of factory-made houses such as metal Lustron homes, new materials such as wallboard or gypsum board for use in traditional wood-framed houses, and laminates perfected in PT boats and airplanes that were then used by furniture designers Charles and Ray Eames and others including Henry P. Glass (fig. 6; see pp. 22–25) all contributed to the explosion of creativity in architecture and design that followed the war.

Certainly given a boost because of the exposure of so many Americans to the German autobahn system during the war, the Federal Highway Act mandated the creation of our own national interstate highway system in 1956. Even before this, however, certain localities funded smaller regional systems (fig. 7) that connected cities to the budding suburbs where new housing was being built and new shopping areas, such as in Park Forest, Illinois (1946–47), were going up. One local firm, Chicago Bridge and Iron Company, patented new designs for water towers such as their water-spheroid tower of 1949 (fig. 8), which it built in countless new communities over the next few decades. The shape of this tower recalls the mushroom cloud explosion of the atomic bomb developed during the war, giving three-dimensional form to an image expressive of the most powerful of technologies, just as other architects and designers adopted clean, curvilinear forms in the designs they created for the atomic era.[6]

Architects such as Mies van der Rohe and his students dreamt of converting some of the large-span factories built during the war into peacetime uses (see pp. 18–21). Wartime factories were reformulated as a matter of survival for their corporate owners. Designers such as Richard Ten Eyck (see pp. 26–29), who worked for Beechcraft, and Henry Dreyfuss, who worked for Convair, developed new peacetime products, including private airplanes for all the trained aviators returning from the war. Dreyfuss himself was one of several designers who, unsuccessfully, proposed combining a car and a plane for the new market (fig. 9). A surplus of large wartime transport aircraft was available for retooling in factories; these ready planes helped jump-start the development of new airports and begin the democratization of the air transport industry well before the jet age of the 1960s.

It is certain, then, that architects and designers used their creative talents and abilities to solve design problems to help win the war. At the war's end, they helped rebuild their respective environments into a new society. What we have briefly outlined in the following pages are but a few of the stories that should remind us of those who lived through those traumatic times and used their wits to survive and provide creative solutions throughout the crisis. Our hope is that future generations will continue to document these contributions and keep alive the story of their heroic and artistic struggle.

Fig. 8 Chicago Bridge and Iron Company. Waterspheroid tanks in Elmhurst, Illinois, 1954.

Fig. 9 Henry Dreyfuss. Convair aerocar, 1946–47.

Bertrand Goldberg and the Legacy of 1945

LORI HANNA BOYER

The year 1945 represents a significant focal point in the career of Bertrand Goldberg and offers a unique opportunity to assess the impact World War II had on his architectural work, especially since Goldberg opened his practice in Chicago in 1937, only two years before the declaration of war in Europe. Like many wartime architects and engineers, Goldberg centered his early work primarily on the development of prefabricated structures. The federal government was interested in industrialization and in the creation of a prefabrication industry that could provide housing in the event of war.[1] But the seeds of prefabrication had been germinating for at least a decade, as the role of architectural design was shifting away from purely aesthetic considerations and becoming more integrated with the creation and advancement of manufacturing processes. Additionally, the 1933–34 Century of Progress Exposition heightened awareness of new industrialized methods of construction and the possibility, for example, of manufacturing a house in a factory for a cost equivalent to that of a car.[2]

Goldberg viewed prefabrication as a process where "the economy of providing shelter became more important than the aesthetics of shelter."[3] They were, after all, "building houses for an industrialized class, not the suburban wealthy."[4] Though a certain restraint was important, Goldberg thought "good design could still exist in a spatially confined shelter."[5] Two of Goldberg's earliest projects that manifested these concepts were the government-sponsored Standard Housing developments in Suitland and Indian Head, Maryland. Goldberg's distinct ability to integrate planning and design skills with engineering and construction techniques successfully realized the systematic production of fifty-seven prefabricated houses manufactured at factories located in Anderson, Indiana, in 1939, and Richmond, Virginia, in 1940.[6] Housing projects such as the Standard Housing developments were regarded as critical to the war effort, and prefabrication became an "essential industry."[7]

Goldberg also applied his talents on behalf of the Board of Economic Warfare under the Office of Strategic Services (O.S.S.) during World War II, designing two mobile structures. For the North African theater, he created a traveling delousing unit to address the threat of typhus fever epidemics among civilian populations (fig. 1). Drawing upon some of his earlier mobile mast-hung designs for an ice cream store and gas station, Goldberg created a convertible truck that could easily be erected in one day. Employing noncritical materials such as plywood and canvas, the truck housed a permanent staff of three, as well as equipment and supplies for a temporary staff of five. The proposed plan of operation involved the fumigation of clothing in the truck body while individuals were showering, but the delousing unit was never realized because the chemical agent DDT was determined to be more effective at killing the typhus-bearing lice.

The second construction design, developed in partnership with Westinghouse Electric and Goldberg's Standard Fabrication Corporation, was for a mobile penicillin laboratory (fig. 2). With the advent of the war, the demand for penicillin increased beyond current production capabilities.[8] Goldberg's structure, made of stressed-skin plywood, facilitated not only the culture and development of the vaccines, but also the packaging and distribution. In a lab design reminiscent of two boxcars placed in a T formation, the penicillin mold could be planted in one section and then grown in a controlled environment within three to four days.[9] By the time the model of this lab was developed, however, the war was nearly over and the demand for penicillin was decreasing. Thus, this project also remained unbuilt.

The economical use of materials was one of the priorities for a wartime government that also encouraged reuse. Goldberg met this demand by designing a convertible shipping crate for a Bofors 90mm antiaircraft gun (see Zukowsky, fig. 5). Once delivered to the field, the crate could easily and effectively be converted into temporary, portable military housing. Goldberg has stated that between 250 and 500 of these crates were built.[10]

Following the war, Goldberg expanded upon principles of prefabrication and systematic development in his architecture and design. He worked from 1949 to 1952, for example, as a consulting architect for the Pressed Steel Car Company, whose president, John I. Snyder, Jr., was interested in making prefabricated railroad freight cars without steel. Building upon the tubular conception of the mobile penicillin lab, Goldberg designed, engineered, and patented stressed-skin plywood and plastic monocoque rail cars,[11] a single-shell technique that had been applied to military uses such as PT boat hulls and some airplanes during World War II.

The name "Unicel" refers to the cellular lamination process that fuses the exterior elements—plywood skin, weathering films, and paint—together with the interior frame and laminated roof, side and end cores.[12] These elements are all visible in a cutaway view (fig. 3). In order to manufacture a freight car of this type, Goldberg adapted the tools, dies,

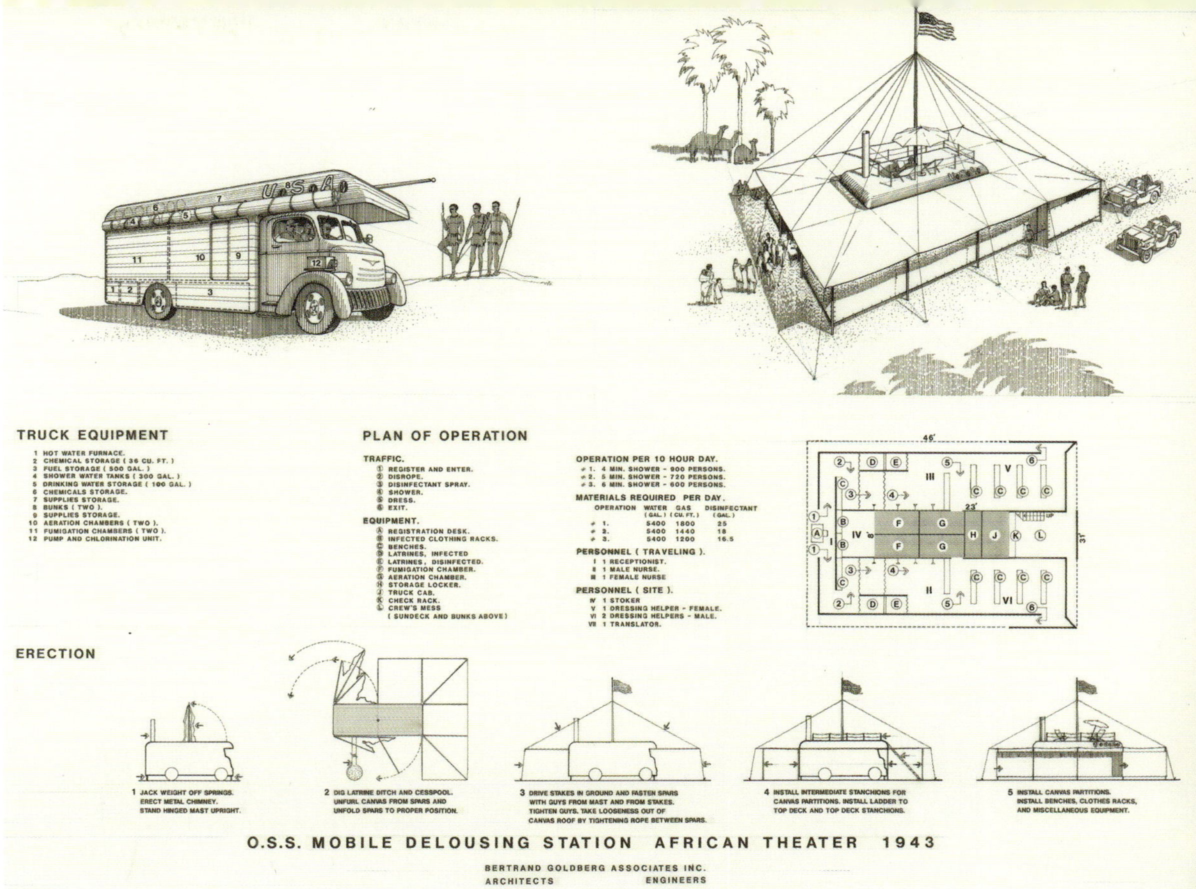

Fig. 1 Bertrand Goldberg. Bird's eye view, plans, sections, and perspective views of a proposed O.S.S. mobile delousing unit for the North African theater, 1943. Blackline on mylar; approx. 76 x 106.5 cm. From the archive of Bertrand Goldberg, gifted by his children through his estate, RX23664/82.2.

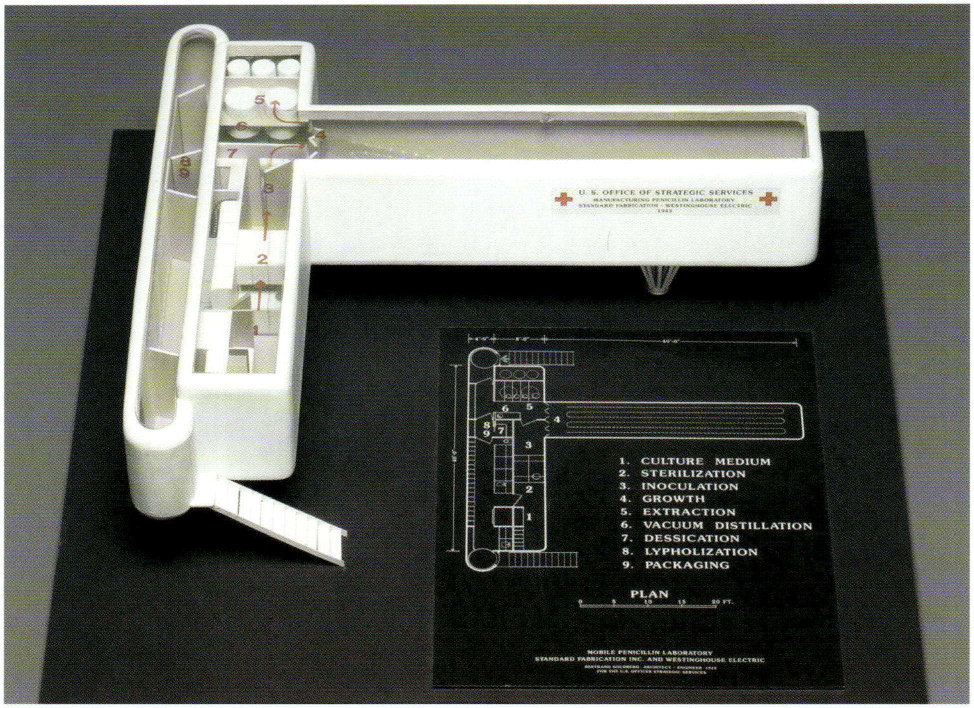

Fig. 2 Bertrand Goldberg. Model of a mobile penicillin laboratory for the O.S.S. and Westinghouse Electric, 1943–44. Mixed media; approx. 50 x 50 x 15 cm. From the archive of Bertrand Goldberg, gifted by his children through his estate, RX23664/158.2.

and press that he had developed for the Stanfab bathroom to make a slab of plywood ten feet wide in any length.[13] This kind of tubular construction, in which individual panels are converted into a single structural tube, was part of a patented invention that advanced commercial manufacturing techniques previously unavailable before and during the war.

Goldberg's refrigerator and ventilator cars met a demand for economy of space, materials, and function. These plywood freight cars—10 feet wide by 10–12 feet high by 40–60 feet long—were amazingly strong and held up well during crash tests because the monolithic design distributed stress throughout the structure, as opposed to conventional steel cars that concentrated six, twelve-foot steel I-beams on the car ends to resist impact.[14] They used "twenty tons less of critical metal than a conventional steel freight car and also served a dual purpose: to carry either perishable food or ordinary cargo. The cars utilized new mechanical refrigeration, instead of traditional ice-cooling, gaining 75 percent more capacity with this type of refrigeration. The car also cost less than standard refrigerator cars and half as much as a conventional boxcar."[15]

The Technical Committee of the American Railroad Association approved the manufacture of 5,000 cars, and 500 were shipped from the factory at Hegewisch, Illinois, to the Arabian-American Oil Company in Saudi Arabia.[16] Although the Army Transportation Corps and the Department of Defense met with plywood industry representatives on October 25, 1951, and seriously looked at the Unicel product as a viable substitute for scarce metals, in the end the Executive Committee of the American Railroad Association rejected it on the basis of the broad ramifications faced by the steel industry if both plywood and lumber were made to supplant metals and other critical materials.[17] When the freight car product failed to enter mass production, Goldberg and his partner, Leland Atwood, again readapted and remarketed their product with Pressed Steel Car Company as Swift Cold Front Express trucks that featured a frameless trailer (fig. 4), and as five-room "Unishelters," prefabricated housing that could be built for less than $8,000.[18]

Goldberg never entirely abandoned the material and economical reutilization strategies he gained while working on wartime prefabrication projects. He continued to create designs independently and then modified them to address the specific requirements of potential clients. Goldberg's vision extended beyond a single architectural discipline to utilize a synergistic approach that combined the fields of architecture, engineering, technology, development, construction, and marketing to present architectural solutions to the economic and social issues faced by urban centers. Although his work during and after the war was similar to projects conducted by many other architects, his unique synchronism can be directly attributed to the techniques he learned during the wartime period.

Fig. 3 Bertrand Goldberg. Cutaway perspective presentation drawing of a refrigerated and ventilated railroad car (Unicel), 1949–50. Mixed media on paper; approx. 76 x 101 cm. From the archive of Bertrand Goldberg, gifted by his children through his estate, RX23664/155.2.

Fig. 4 Bertrand Goldberg. Perspective views of a Swift Cold Front Express Trailer for Pressed Steel Car Company, c. 1949–50. Colored, printed paper; approx. 52 x 67 cm. From the archive of Bertrand Goldberg, gifted by his children through his estate, RX23664/155.1.

Improvisation and Inventiveness in Bruce Goff's Quonset Designs

CARISSA KOWALSKI

As a young man, Bruce Goff was drawn to the work of Frank Lloyd Wright, but Goff soon realized he needed to find his own architectural approach. His interests in painting, music, and even industrial products helped inform this search, but it wasn't until World War II that Goff was given the opportunity to begin the quest in earnest.[1] His service in the United States Navy gave him time to contemplate design values and priorities and to cultivate his own way of solving aesthetic problems. Goff's experiences in the 1940s—both as a military man and a private citizen—allowed him to explore new materials and forms that greatly influenced his life's work.

The limited materials and restrictions imposed by the war encouraged—rather than inhibited—Goff's inventiveness. According to him, "any materials were fair, old or new . . . we had to reevaluate them to use them in ways that seem fresh and valid."[2] One of the materials that appealed to Goff was the Quonset hut, which he first used out of necessity, but as his postwar career advanced, these prefabricated structures became the source of formal experimentation. Goff's emphasis on color, texture, and unconventional building materials was unusual among modern architects during this period, and, although he frequently used industrial materials, byproducts, and prefabricated components, Goff's aesthetic was not industrial. Rather than subscribing to the International Style or any form of postwar conformism, he developed a series of projects that were expressionistic, complex, and unique.

Goff was working in a private practice in Chicago when he enlisted in the Navy in the summer of 1942 and joined the Seabees—a name that came from the initial letters of their official name, the Construction Battalions. As one of the more than one thousand architects involved with the Seabees during World War II, Goff was responsible for a number of projects on military bases stateside as well as several unofficial projects for his commanding officers.[3] He began his first eighteen-month posting in the Aleutian Islands in September 1942. The Alaskan tundra was stark and unforgiving, offering little in the way of stimulation, yet Goff would later recall these months as a time spent "just thinking about architecture and the principles of design."[4] This period of contemplation had a maturing influence on Goff and allowed him to distill his thoughts about color, proportion, harmony, geometry, and texture into a more cohesive design approach.

It is likely that the Aleutians offered Goff his first experience with Quonset huts and their applications as housing, storage, and other naval facilities. The Quonset hut was developed in 1941 by a team of naval architects and soon became the ubiquitous building type of military installations at home and abroad. Because Quonset huts were lightweight, stackable, and easy to assemble, around 170,000 were produced during World War II. Over eighty-six different plans for Quonset structures circulated among military construction crews, from housing and mess halls to hospitals and hangars.[5] Stran-Steel, one of the largest manufacturers of Quonset huts during the 1940s, anticipated widespread civilian usage of their products when the war was over.[6] While these predictions fell significantly short, some architects—like Goff—were intrigued by the possibilities of this prefabricated structure for low-cost housing and other domestic uses.

In March 1944, Goff was transferred to Camp Parks, a base located about thirty miles southeast of San Francisco that was used by the Seabees as a place to recuperate, train, and reorganize between missions in the Pacific. The buildings, therefore, were not as combat-oriented or sterile as those found in other bases in the United States or abroad; they were intended to increase the troops' morale.[7] Within the more-or-less bureaucratic regulations of the Navy and the inevitable

Fig. 1 Bruce Goff. Interior perspective view of Quonset hut chapel and library, second proposal, Camp Parks, California, 1944 (later moved to San Lorenzo, 1947). Graphite on tracing paper; 73 x 85.2 cm. Gift of Shin'enKan, Inc., 1990.855.

Fig. 2 Bruce Goff. Elevations of proposed Stran-Steel Quonset houses, 1945. Graphite on tracing paper; approx. 38.2 x 36.3 cm. Gift of Shin'enKan, Inc., 1990.819.13.

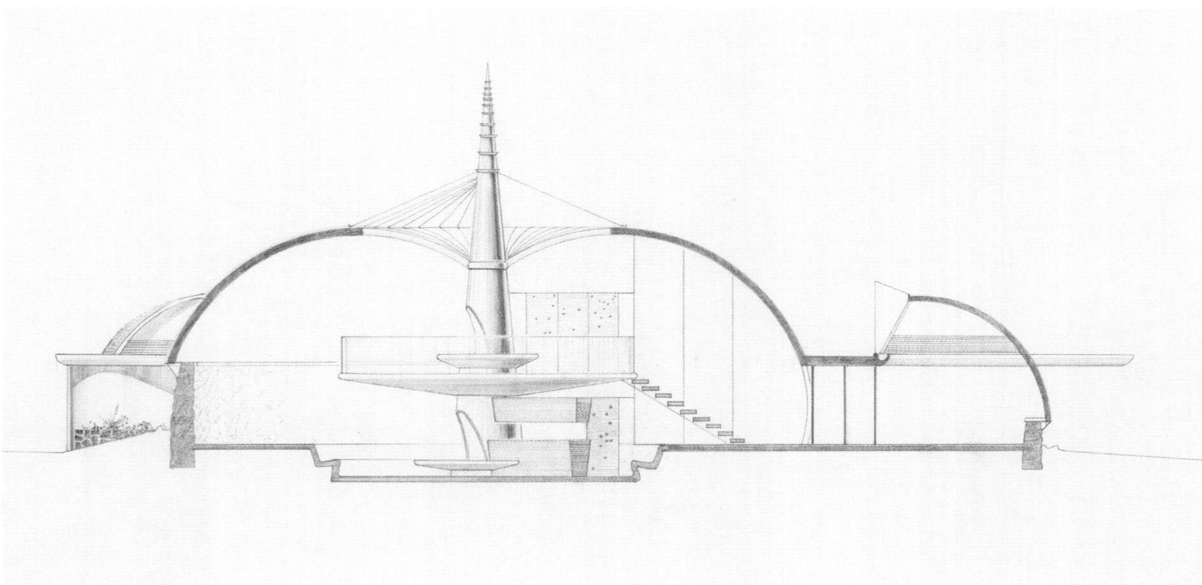

wartime material shortages, Goff was able to create structures that used "official" materials in a creative and improvisatory way. Most of Goff's designs for Camp Parks were renovations rather than new construction, including work for the base's Hostess House, Officers' Club, and Ship's Service Store. Goff's only ground-up project was the McGann Memorial Chapel (commonly known as the Seabee Chapel), built during the last days of the war (fig. 1).[8] Goff's first design for this multifaith chapel called for laminated wood framing, but when surplus Quonset huts were made available at Camp Parks in 1945, Goff redesigned the chapel around this new building system.

Two "elephant" Quonsets (so named for their oversize layout) were arranged end-to-end to form the body of the Seabee Chapel. To this basic shell, Goff added a variety of materials of contrasting colors and textures. Salmon-colored brick pylons break up the horizontal space on both the exterior and interior, and landscaping elements, including a reflecting pool, modify the usually stark Quonset structure. Goff softened the long, narrow interior space by adding a clerestory element and rows of hanging plants in round glass bowls. The corrugated metal cladding of the Quonset huts was juxtaposed with unstruck mortar joints and random rows of projecting brick. Goff's interest in using indigenous materials is apparent in his choice of redwood for the pews and fascia boards.[9]

After the war ended, Goff remained in California to open a private practice. He continued with the material improvisation that characterized his Navy career, and in 1945 he developed a series of small houses using Stran-Steel Quonsets as a low-cost starting point for the design (fig. 2). The variations on this theme used two short lengths of Quonset huts of varying heights joined by a flat-roofed circulation space and capped with a combination of wood, masonry, and stone. These residences (none of which was built) are similar to other Goff houses in that they are composed of geometric elements, an open plan, and juxtaposed materials of varying textures.

As he experimented with Quonset forms, Goff wandered farther away from the linear, "gun-barrel" structure. For him, it was not just a matter of substituting new materials into a framework of traditional house design; instead, Goff examined new forms that began with the possibilities the materials themselves suggested. Goff's Quonset designs reached a pinnacle in 1947, when he returned to the Chicago area for a commission for Ruth Ford, a friend of Goff's and the director of the Chicago Academy of Fine Arts.[10] The Ford family needed space for living, entertaining, and exhibiting Ruth's art, and Goff designed a round, central space with three levels for these functions and two quarter-circular "satellite" spaces that contained the bedrooms and bathrooms (figs. 3–4). Goff's use of found and military surplus materials is evident in many aspects of the Ford House's design: Quonset ribs, for example, deconstructed and painted red, formed the basic structure that radiated out from the central mast. The low exterior walls were composed of black cannel coal and green glass cullets, with mortar joints studded with colored marbles. Goff used Navy surplus rope for the flat ceilings and soffits, and he employed Plexiglas "Flying Fortress" domes as skylights in the bathrooms.[11] This rich palette of materials combined to create a house that was not only worthy of its artistic client but also a local landmark.[12]

Goff was inevitably shaped by the decade of the 1940s, but only to the extent that his experiences strengthened and crystallized his approach to architectural design. His development, as seen through three of his Quonset hut–inspired projects, suggests a designer coming into his own and exhibiting a creativity that would make him a prolific architect well into the latter half of the twentieth century.

Figs. 3–4 Bruce Goff. Elevation and section of the Ford House, 404 South Edgelawn Avenue, Aurora, Illinois, 1947–49. Ink on tracing paper; 48.4 x 91.2 cm.; 44.4 x 86.2 cm. Gift of Shin'enKan, Inc., 1990.809.13 and 1990.809.15.

Ludwig Mies van der Rohe and Chicago in the 1940s

MARTA WOJCIK

Ludwig Mies van der Rohe's first visit to Chicago in 1937, though just a short episode, was a prelude to his long-term commitment to the city's architecture. It was then that he first saw the City of Big Shoulders and the tall commercial buildings that exemplify the Chicago School of Architecture. Whatever he thought of the formal aspects of these buildings, he must have liked the way Chicago architects and engineers used steel, because he himself favored this material in his American practice. Thus, it is no coincidence that he became the father of the so-called second Chicago School of Architecture. It was also on this first trip that he paid a visit to Frank Lloyd Wright at Taliesin in Wisconsin. Their careers went in considerably different directions, yet Mies nonetheless highly regarded Wright's approach to architecture, especially the spatial openness of his projects.

And, finally, on this trip to America Mies had the opportunity to meet with representatives of the Armour Institute of Technology. Mies was a candidate for the directorship of the School of Architecture at Armour, which very soon became the Illinois Institute of Technology as the result of a merger of Armour Institute and Lewis College. The committee was searching for an eminent individual who would be able to revitalize the school's academic program, and Mies ultimately accepted their offer in 1938 and moved to Chicago the very same year, a decision that may have been informed by Mies's increasing feelings of constraint as an artist and educator in Nazi Germany. The frustration he shared with many other German citizens was expressed a couple of years later in a bitter comment on the situation after his homeland was destroyed by the Allies: "You can't have an animal like Hitler loose in the world, and if it means annihilating Germany in order to accomplish that, then that's what has to be done."[1]

Even though Mies hardly spoke English and knew little about the city, Chicago was his home for the rest of his life. Along with his German-émigré colleagues Ludwig Hilberseimer and Walter Peterhans, Mies developed a progressive new curriculum for Armour's School of Architecture based on Bauhaus methods. As one of the students recollected: "I think IIT was a fantastic school, the curriculum that Mies set up, because you learned all these things— how wood goes together, how you build with steel, concrete, brick, masonry, all these materials. . . . I think there was no other school in the country that was comparable in the grounding of fundamentals."[2] Apart from creating the new curriculum, Mies also undertook the challenge of designing the master plan for the new IIT campus, which in his own words was "the biggest decision [he] ever had to make."[3] By the time Mies arrived back in Chicago, Armour's

School of Architecture was housed in the Art Institute of Chicago and, like the other Armour departments, it was in desperate need of a new space. Mies's proposal led to the first modern campus realized for an American university.

In some ways, the decision Mies made in 1938 did not pay him back right away. At first, he could not proceed with his ideas, for it was expected that most, if not all, construction should be oriented toward the war effort. Thus, most of the IIT buildings Mies designed were only constructed after 1945, while some, such as his proposal for an Administrative Building, would never be realized (fig. 1). The only exception was the laboratories of the Armour Research Foundation, which focused on war-related research during World War II.

During the war Mies's students were coming and going, their education interrupted by military service. Some of them, however, had a chance to develop their architectural skills while serving their country. This was the case, for example, with Myron Goldsmith, who was appointed to the Army Corps of Engineers and who, after the war, utilized this experience in his practice at Skidmore, Owings, and Merrill.[4] Mies's design for his IIT buildings differed from his later American oeuvre; on the campus he decided to use brick extensively. In some part this was because of the budgetary limitations as well as the scarcity of steel during the war.

The construction of the first academic building on campus was, not surprisingly, the Naval Science Building, which began in 1945 and was meant to serve as an armory.[5] After the war, when the name was changed to Alumni Memorial Hall (in recognition of alumni who had died in service), it was taken over by the Naval Reserve Officer Training Corps and the School of Architecture, the latter gaining its own quarters in Crown Hall a decade later. This building was in some ways the manifestation of Mies's philosophy, according to which the structure should be clearly expressed. He exposed the I-beams welded to the fireproofed structural columns at the corners of the building to articulate the essence of the structure's steel skeleton. Throughout the war Mies worked tirelessly on his vision for postwar architecture, exploring two ideas that would fascinate him throughout his American career: the high-rise and clear-span construction.

In 1942 Mies and his students were working on a conceptual project for a concert hall (fig. 2), a design that had a profound influence on his ideas for

Fig. 1 Ludwig Mies van der Rohe, with the assistance of George Danforth. Rendered elevation of an early study for Administration Building, Illinois Institute of Technology, c. 1941. Pencil and charcoal on Strathmore paper; 78.4 x 101.8 cm. Gift of A. James Speyer, 1988.31.2.

Fig. 2 Daniel Brenner. Graduate student project at Illinois Institute of Technology for Ludwig Mies van der Rohe, 1946. Perspective view of proposed Theater and Arts Complex within an Existing Space; Study of the Interrelation of Space and Finish Materials, based on Mies van der Rohe's 1942 Concert Hall Project. Collage over photograph of Albert Kahn's Martin Aircraft Factory (1937). Paper, foil, and wood veneer on photo enlargement, mounted on board; 36 x 74.2 cm. Gift of Rachel, Jon, and Ariel Brenner, 1988.33.

later clear-span pavilions. The concept was inspired by a photograph of Albert Kahn's 1937 aircraft factory, an industrial large-span structure that provided a great example of the open space that could be easily divided by freestanding walls or hanging partitions. Mies was fascinated with American building methods and the economical language of technology applied in utilitarian buildings. As he said, "Wherever technology reaches its real fulfillment, it transcends into architecture."[6] For Mies, Kahn's industrial plants were the symbols of their time and he described them as "the cathedrals of the Twentieth Century."[7]

By 1943 Mies and his IIT students were studying the expression of high-rise designs, continuing an idea that he had conceived as early as 1920, when he undertook the Friedrichstrasse Office Building and Glass Skyscraper projects (both unrealized). Mies's primary concept of reinforced concrete and steel structures was explored by Earl Bluestein, Joseph Fujikawa, and other students.[8] Thus, the preliminary designs for both Promontory Apartments (1946–49) and the 860–80 North Lake Shore Drive Apartments (1948–51) were conceived well before Herbert Greenwald, a young developer, approached Mies in 1946 with the offer to invest in the projects (fig. 3). Until that moment, Mies's high-rises, as well as most of the IIT buildings, existed either only on paper or as three-dimensional models. It seems as if through their restless creative activity, Mies and his colleagues strove to "make an order out of the desperate confusion of [their] times."[9]

The erection of the Promontory Apartments in reinforced concrete immediately after the war marked the transitional period when steel was still not obtainable. The construction of the 860–80 North Lake Shore Drive Apartments in 1949 was a milestone that crowned Mies's efforts to express the high-rise structure, a problem he had worked on since the early 1920s. Using prefabricated steel sections, Mies approached his aesthetic goal to elevate technology to the status of art and launched a new era of architecture. Throughout the years Mies trained and collaborated with many talented people including Bertrand Goldberg, A. James Speyer, George Danforth, and Daniel Brenner. His dedication to his principles tremendously affected these young disciples and inspired them to search persistently for self-expression, certainly a lasting part of Mies's legacy to Chicago.

Fig. 3 Ludwig Mies van der Rohe. Perspective sketch of high-rise buildings, possibly Promontory Apartments or 860–80 North Lake Shore Drive, c. 1946–48. Graphite on note paper; 15.2 x 21.7 cm. Restricted gift of the Architecture Society and the Alexander C. and Tillie S. Speyer Foundation in honor of John Zukowsky; through prior gift of the Three Oaks Wrecking Company and Carson Pirie Scott and Company; Samuel P. Avery Endowment and Edward E. Ayer Endowment in memory of Charles L. Hutchinson, 1992.9.

Henry P. Glass and the Beauty of Manmade Things

KAY MANNING

Fig. 1 Henry P. Glass. Advertisement for wheeled shopping cart, part of the "Victory" series for W. L. Stensgaard and Associates, Inc., c. 1943.

Freed of Nazi tyranny in Europe and emboldened by quick success in his adopted home of America, Henry P. Glass flourished in the 1940s, arguably the most productive decade of his almost seventy-year career in industrial design. He followed the artistic sensibilities he learned in his native Vienna, turning out beautifully rendered drawings of everything from radios to collapsible chairs, while always adhering to his own rules for good design: proper use of materials, disdain for senseless ornamentation, attention to aesthetics, and the all-important emphasis on function.

Glass, who was imprisoned in two German concentration camps, immigrated to New York with his wife, Eleanor, in 1939. Here he was deferred from military service because "my draft board decided I was doing more important work for the war effort as a civilian."[1] Though trained as an architect in Austria, Glass never had a U.S. license, leaving him to concentrate on designing products that both capitalized on materials developed for the war effort and anticipated the needs of a postwar society.

While in New York, Glass worked for noted designer Gilbert Rohde and helped him with the architecture and interiors of a pavilion at the 1939 World's Fair. Glass moved on to architect Morris Sanders, whose friend, designer Russel Wright, was trying to create a new look with his American Way products. However, the economy faltered and Glass was let go. In 1942 the Chicago display firm of W. L. Stensgaard and Associates, Inc., asked him to be its director of architectural design, which marked the beginning of his concentration on furniture innovation. In all, Glass was awarded more than fifty patents, many for collapsing and folding mechanisms for chairs and tables.

Functionalism, the theory that beauty in useful objects is defined by their utility and honesty to materials and structure, was the aesthetic standard of the time. Its vocabulary included geometric forms with simple silhouettes, smooth finishes that obliterated signs of manufacturing, and no applied ornamentation. "Styling," or the altering of the exterior of an object without changing its function, was discouraged. But as the war wound down, and in the face of accelerating technology and expendable possessions, this long-established standard became more relative.[2] Glass's "Victory" series for Stensgaard, for example, produced a shopping wagon for carting groceries (fig. 1) and a sturdy utility chair of bent Masonite that was depicted in advertisements as strong enough to support three adult men (fig. 2). His 1941 coffee tables were simple and graceful, with slabs of glass resting on S-shaped bases of bent plywood (fig. 3). In the same year, he designed cut-out chairs of plywood in two versions—the arms curving forward to form the front straight legs or backward to create the rear legs, braced with a runner on the floor (fig. 4). His 1942 "hairpin" chairs and tables, inspired by his wife's hair accessories, featured legs of narrow steel rods, an innovative use of the metal.

Glass approached the expected postwar demand for housing and smaller living spaces with plans for a trailer system in which bathroom, bedroom, kitchen, and living room units were manufactured, then offloaded and sealed together at the home site. As part of the Blue Sky Project in 1944, he designed a dinette table on rollers that could be pulled out for meals. When rolled back, it served as a countertop in a very compact kitchen.

In 1945 and 1946, under the auspices of an architectural firm, he designed new offices for Vincent Kling Studios on Fairbanks Court in Chicago. He used huge window panes and swinging glass doors to create a dramatic entrance and contrasted vertical planks with a curvilinear walkway under clever lighting to point visitors to the receptionist in the lobby.

In the mid-1940s, radios went from being a piece of furniture to a tabletop accessory, opening the door for more streamlined, sound-effective cabinets. His 1946 design for FADA was one of many for various manufacturers in which he experimented with speaker baffles, channel dials, and knobs in cases shaped from upside-down triangles to boxy squares. As television grew in popularity, he designed stands that swiveled and tilted, all in an effort, he wrote in promotional materials, to accommodate children on the floor or adults on couches.

Glass often was his own best salesman, personally presenting sketches to prospective clients from the design studio he began in 1946. He would leave nothing to chance, printing on the rendering what he considered the product's best feature or practical use, and emphasizing it with an exclamation point.[3] He even made palm-size scale models of some of his collapsible pieces, believing if the models caused problems, so would the full-size versions.

Also in 1946, Glass convinced the School of the Art Institute of Chicago to

Fig. 2 Henry P. Glass. Advertisement for utility chair in the Structural Bends furniture line, part of the "Victory" series for W. L. Stensgaard and Associates, Inc., 1943.

establish an industrial design course, which he taught for more than twenty years. His teaching and mentoring are considered an important part of his legacy, as is his book, *The Shape of Manmade Things*, in which he used many of his own creations as illustrations after first grounding his readers in historical precepts.

Glass wrote that good design, "in contrast to good music, good literature, good food, good fun or even good art,

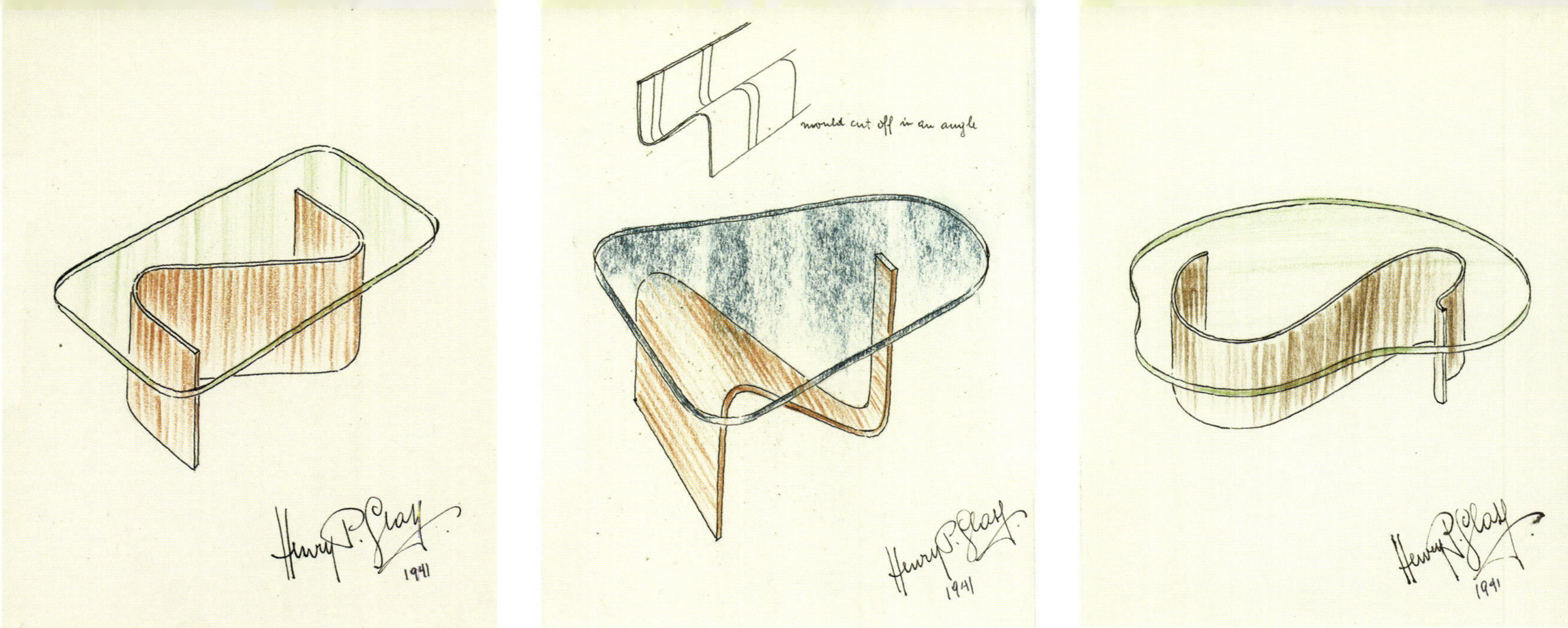

Fig. 3 Henry P. Glass. Perspective drawing of glass and bent plywood table designs, 1941. Pencil and colored pencil on sketch paper; each approx. 18 x 13 cm. Gift of Henry P. Glass, 1999.547.9–11.

which are all subject to personal taste, style, fashion or fad, is governed by indisputable, eternal rules, unalterable by conditions of historic environment or location."[4] He considered it essential for designers to keep current with technological advances (although he never mastered a computer), especially in their choice of materials. He acknowledged a certain consumer preference for natural materials, but praised the utility of plastics engineered in the mid-1940s, and he wrote that "if the new technology is allowed to be honestly applied, it assumes a beauty all its own, without fakery."[5]

As the demand for consumer goods exploded across the world, Glass responded with evermore creative designs for furniture, glassware, cutlery, lamps, and more. He adapted the science of ergonomics, which emerged as a discipline during the war, to many of his furniture pieces. His fascination with knock-down chairs and tables led to two of his more successful designs—a folding conference table for Samsonite and a tubular aluminum Cricket chair that folded to a thickness of one inch. In 1960 he drew a foldable aluminum "mobile living space" for travel and camping, called the Accordium, based on the principle of bellows.

Glass's original training as an architect spurred him to draw hotels, vacation homes, and apartments, inspired in part by R. Buckminster Fuller, whom he revered as "the greatest design innovator of our century, a poet of technology."[6] Glass's Quadrangle Clusters, groups of factory-made modules on stilts that allowed for parking beneath, were drawn for the Mobile Homes Manufacturing Association and were an evolution of his earlier trailer system. His Coast Castle for the Douglas Plywood Company, comprised of rectangular glass cubes, was, in Glass's words, "a dream place in Hawaii or the Virgin Islands where a family could spend a vacation out of this world."

Glass so believed in Fuller's Dymaxion principle of harnessing nature and technology for maximum efficiency that he hoped to acquire one of the designer's prefabricated "dwelling machines" for his personal home, but his deposit was returned because the speculative project was abandoned after the war.[7] Glass then designed a passive solar house and moved his family and studio to Northfield, Illinois, where he toiled almost until his death at age ninety-two in 2003.

In all of his patents and enduring designs for collapsible chairs—as well as his Swingline furniture, for which he won the coveted Industrial Design Institute Gold Medal—Glass never quite achieved the fame he might have if he'd worked on the East Coast, had flashier clients, or gone against his firm beliefs about the role of an industrial designer and the creative process. He was considered by some in the field to be very farsighted, maybe too much so for the average consumer's taste at the time. His color-coded toy chests with swing-out drawers for Fleetwood Furniture Company (fig. 5) in the early 1950s are an example of how he was ahead of the curve. He said of his children's designs for Swingline, which he called a commercial failure: "Originality might have been part of the reason. . . . All this was too new and startling and could not overcome the preference for the customary, old and proven market offerings."[8]

Half a century later, Glass's ingenuity is recognized with the possible reintroduction of two of his designs: a Swingline-style cabinet, suitable for adults or children, and a folding table with special spring mechanisms that enable the use of both sides of the tabletop.

Glass was unrelenting in his pursuit of good design, and in a speech to upholsterers he offers what could be a coda to his career: conformity, shyness about scientific advances, lack of originality, surrender to commercialism, and a reliance on reactionary decoration are "mortal sins when judged by standards of ultimate human values. Never, ever should the designer stop studying the increasing and expanding resources at his disposal. . . . If he does, he loses his values to society and to his clients as well."[9]

Fig. 4 Henry P. Glass. Perspective drawing of molded plywood chair, 1941. Pencil and colored pencil on sketch paper; approx. 23 x 16 cm. Gift of Henry P. Glass, 1999.548.1.

Fig. 5 Henry P. Glass. Presentation drawing for a toy chest, Swingline Juvenile Furniture, Fleetwood Furniture Company, c. 1950–52. Colored pencil, pastel, and graphite on tracing paper; approx. 33 x 35.2 cm. Gift of Henry P. Glass, 1999.543.1.

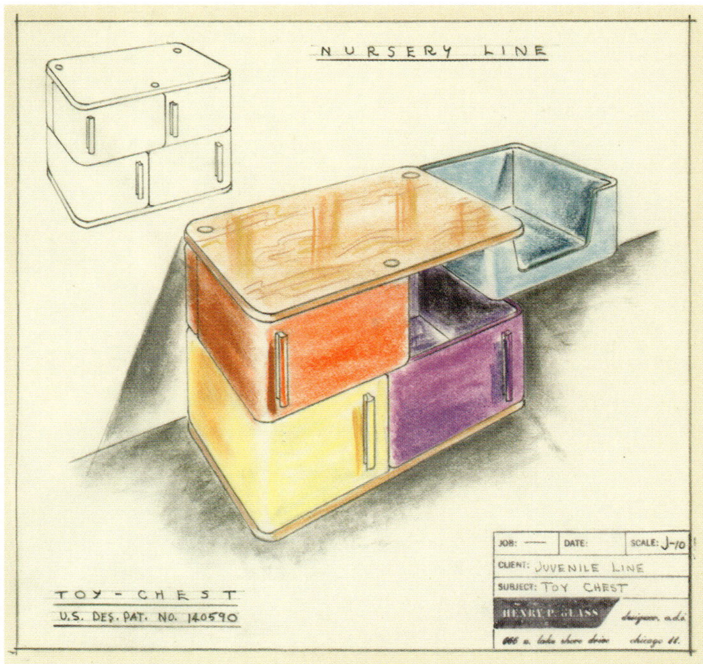

Richard Ten Eyck and Industrial Design of the 1940s

MARTA WOJCIK

When Richard Ten Eyck began to study industrial design at the University of Illinois in 1938, the academic base for this field was hardly developed.[1] Most of the American pioneers in industrial design adopted this profession after they had completed their formal education in related fields. Many were graduates in architecture who turned to industrial design after their futile efforts to find a job during the Depression. Ten Eyck's interest was first turned to the visual aspects of highly functional products. Forced to leave the school in 1940 due to financial and health problems, he found employment as an industrial designer for Pneumatic Tool Company (later Thor Tool Company) in Aurora, Illinois, where he designed an innovative line of electric hand tools.

In 1943 Ten Eyck joined Dave Chapman, Inc., where he worked on products for Montgomery Ward. Dave Chapman was a successful Chicago-based designer with an architectural background. Through the experience that he gained working for the 1933–34 Century of Progress Exposition and later as a head of product design for Montgomery Ward, Chapman had decided to open his own office in 1935. During World War II Chapman's firm focused on adapting products to materials that were not under wartime control, at a time when designers had to come up with new economic solutions because of the scarcity of traditional materials. As noted furniture designer George Nelson wrote in 1943: "After decades of functionalist preaching . . . we turn unquestionably to functional design."[2]

In 1944 Richard Ten Eyck left the Chapman office to work on drafting tool designs for a new military airplane. This first experience with the aerospace industry would soon become his lifelong fascination. Toward the end of the war in 1945, Ten Eyck moved to Wichita, Kansas, where he accepted a job at the Beech Aircraft Company. He was responsible for the interior design of a new airplane, the Model 35 (also known as the Bonanza; fig. 1) and an experimental car called the Planesman. The Beech Bonanza was built for civilian flyers and was first flown in 1947. This kind of plane was promoted as a new alternative to the family car, but the high cost of this enterprise precluded any realization of such a dream. Nonetheless, the Bonanza found many enthusiasts, and twenty years later there were 10,000 Bonanzas in service, and improved versions were still in production.[3] Ten Eyck recounted his experience at Beech Aircraft Company as an important period when he learned design principles that were critical factors, not only for consumer comfort, but principally for safety. While he was working at Beech, Ten Eyck also conceived of a design for the Vornado fan for the O. S. Sutton Corporation, a product that

Fig. 1 Congressman Peter F. Mack, Jr., in a Model 35 Bonanza at Beech Field, April 19, 1952. Interior designed by Richard Ten Eyck. Manufactured by Beech Aircraft Corporation.

demonstrates a clear relationship to aeronautical design of the era (fig. 2).[4] Ten Eyck originally styled the fans to look "like the front end of an airplane."[5] Indeed, his creation was consciously inspired by the prop-engine nacelle of the airplanes first used during the war, and unlike traditional electric fans, the Vornado pushed air in a concentrated cone instead of dispersing it.[6] Ten Eyck also applied a "dust-proof gray" color that additionally distinguished the "chubby metal-cowled machine" from typical fans of the era.

Ten Eyck came back to his 1940s design of the Vornado fan when in the 1980s he was asked to update the look without losing the original streamlined shape. The original design, however, was changed rather considerably (fig. 3). In an article from the late 1980s, the author commented on the growing interest in postwar design: "The first Vornadoes, 'a product of the mid-twentieth century, the streamlined age,' the ads told us, are today sought after by collectors of all those toasters, irons, and vacuum cleaners that look as if they were designed in a wind tunnel. Only having a Vornado is like having a wind tunnel itself."[7]

Ten Eyck continued to work for Beech Aircraft through the end of the war. But in the days that followed V-J Day, victory celebrations were overshadowed by the threat of unemployment, a consequence of the closing of thousands of the enterprises oriented toward war production. Ten Eyck recalled the day after Japan surrendered to the United States on August 14, 1945: "I stood across the street from the main gate and there was a line of people being terminated that was a block long. . . . I was just about ready to cry, thinking about all those people out of work, that we were going back to the damned depression again." Yet there was also a considerable dearth of consumer goods, because the war depleted many products needed for everyday life. Ten Eyck thought he might respond to both problems by forming a business, and he did so in 1948, heading it until his retirement in 1995.

The first project designed in 1948 under the name of Richard Ten Eyck Associates was a new version of a sandwich shop manufactured by Valentine Industries in Wichita (fig. 4). The models sold by Valentine received a streamlined form and featured a bar of stainless steel, colored Formica or Vitrolite countertops with stainless steel front and trim, electric wiring, electrical refrigeration, a drive-in, and a front entrance of stainless construction. The company praised the model for its efficient, sanitary operation and the attractiveness of its design.

Fig. 2 Richard Ten Eyck. Sketch for a Vornado fan for Kenmore, 1949. Pencil and graphite on tracing paper; 30 x 21 cm. Gift of Richard Ten Eyck, RX21677.

Fig. 3 Richard Ten Eyck. Vornado fans, past and present.

The product was inexpensive and could be operated by just one or two people.[8]

In the late 1940s Ten Eyck became acquainted with the work of the Cessna Aircraft Company. Small projects led him and his firm to a long-term design consultancy for Cessna (1950–80), where he worked on the exterior of the Cessna 172, once the most popular single-engine airplane in the world.

Richard Ten Eyck belongs to the generation of inventors who, despite the world crisis, managed to face dramatic problems and realize innovative and beautiful ideas. Ten Eyck successfully launched his career in the 1940s and continued to be a leader in the field until 1995. Working on a wide range of products (figs. 5–6) he believed that the design of the final object "should convey everything the designer wants it to say to the customer."[9] The design should reveal the truth of the product, provide the user with confidence, and ultimately provide aesthetic pleasure. After all, industrial design is art, as is proclaimed in the first publication of the Society of Industrial Designers: "The industrial designer realizes that basically he is an artist, although he may not stress the fact. . . . While he inevitably gives the stamp of his own personality to his work, he is not trying to express either his personality or his emotions. He tries, rather, to organize any object or structure in the form best suited to it, so that it will give the maximum of pleasure and satisfaction to its ultimate user." As a creative artist, Richard Ten Eyck responded ingeniously to the needs of the era, a fact reflected in the long-term cooperation he established with numerous clients and the positive response the public gave to his designs. The 1940s were filled with many experiences crucial for Ten Eyck and his long and successful career as an industrial designer.

Fig. 4 Richard Ten Eyck. Perspective design study of sandwich shop for Valentine Industries, c. 1945–48. Pencil on tracing paper; approx. 43 x 35 cm. Gift of Richard Ten Eyck, RX21677.

Figs. 5–6 Richard Ten Eyck. Exterior and interior studies for a "Lighthouse" double-decker trailer, c. 1950–53. Pencil on tracing paper; each approx. 56 x 46 cm. Gift of Richard Ten Eyck, RX21677.

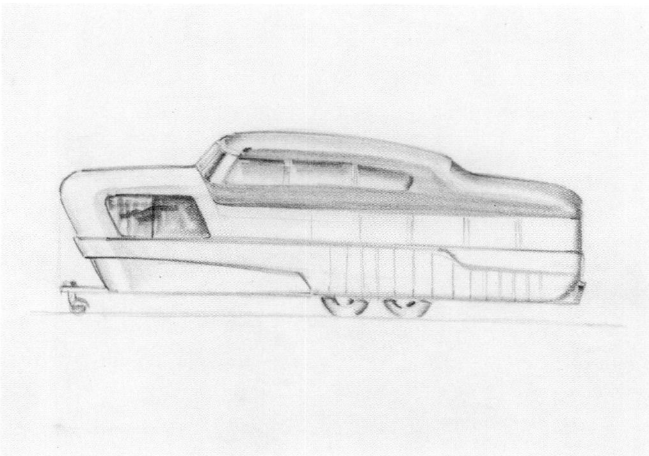

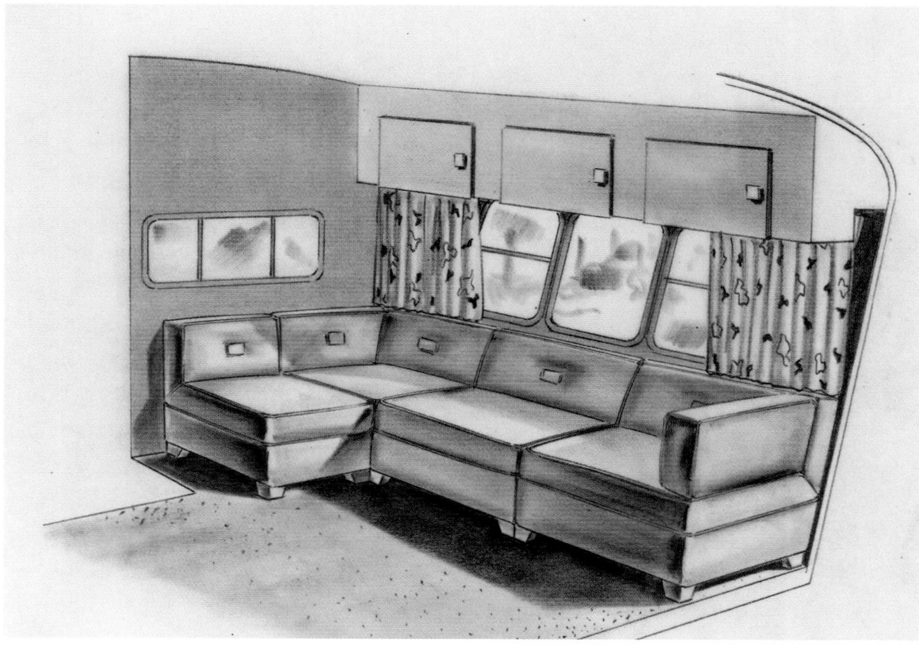

L. Morgan Yost's Houses for Modern Living

MARTHA THORNE

L. Morgan Yost is perhaps best remembered as an architect of single-family homes, most of which are located in the northern suburbs of Chicago. These are not, however, opulent recreations of historic styles; rather, all of his residences could be called "modern": straightforward designs that thoughtfully respond to the natural characteristics of the site and seek efficiency and comfort at a reasonable price. A native of Ohio and a 1931 graduate of Ohio State University, Yost established his own practice in the Chicago suburb of Wilmette in 1933. During his first years in business, with the country suffering through the Great Depression, most of his commissions consisted of smaller-size remodeling projects or furniture designs. In time the economy improved and his practice grew, yet Yost contracted tuberculosis in 1938, which required his extended stay at a Naperville sanatorium and months of recuperation. Apparently the six months he spent in the hospital gave the architect time to develop his ideas about architecture and professional practice, and he published an essay in which he lamented that the practice of architecture had become more a business and less a genuine and personal expression of the architect.[1] He asserted that the distinguishing characteristic of the new American home was that it was *functional*, not only in terms of structure, but also in terms of its ability to provide comfort and respond to the demands of modern life.

One of Yost's first designs to embody his idea of function and balance in architecture was his own home and studio in Kenilworth, Illinois. Widely published and even shown in an Art Institute exhibition in 1942,[2] the design attracted interest on the part of architects and the general public and reflected his attempts to create an ideal environment for all the varied physical activities carried on in the house. Spaces were arranged for efficiency and elements were incorporated to promote natural light and comfort. In all of his designs, Yost stressed the importance of sensitive siting—using the natural attributes of the site to guide the design and make the house and landscape part of a single conception.

During World War II, with little architectural work available and disqualified from service by his illness, Yost looked for other opportunities in the field. He worked as editor of *American Lumberman,* a magazine directed toward retail lumber and building materials dealers. He also taught at the School of the Art Institute of Chicago, edited *Small Homes Guide*, a semi-annual publication of the National Homebuilders Bureau, and served as consulting editor to *Household* magazine. As the country hoped for the end of the war, a new definition of the American home was evolving, and modern architecture was part of that new concept.

Fig. 1 L. Morgan Yost. Aerial perspective of the Zoned House from "House Studies," published in *Small Homes Guide* (Fall 1944). Ink and colored pencil on tracing paper; 21 x 34.6 cm. Gift of L. Morgan Yost, 1985.583a.

Fig. 2 L. Morgan Yost. Aerial perspective of Homes for Our Children from "House Studies," published in *Small Homes Guide* (Spring 1944). Ink and colored pencil on tracing paper; 31.8 x 45.6 cm. Gift of L. Morgan Yost, 1985.577a.

Many publications, especially those of the type that Yost edited, had an optimistic air, looking forward to returning GIs, new development in the suburbs, and the potential for new inventions and technology. It was common for such publications to show a variety of new homes, complete with plans that could be obtained by the public. This was a way to reinforce the idea that a new home would be accessible to all. Even the influential journal *Architectural Forum* clearly stated in 1942 that one of its objectives was to dramatize the contributions of modern design to building a better America: "The small house is . . . in a good many ways . . . what we are fighting for."

Yost's works—those built in the Chicago area and those that existed only on paper as proposals or demonstration projects—have recurring themes of efficiency, comfort, and functionality, all for a reasonable cost (figs. 1–2). The Deno House (completed in 1947) and the Erskine Residence (completed in 1946), both in Highland Park, and the Morse House (1945) that he designed for Kenilworth, Illinois, all emphasize his ongoing concerns. In every case, the site and house work together. If the ground on a given plot was sloped, Yost used the natural topography to erect a split-level house, allowing abundant natural light into the interior, while separating the different types of functions on different levels. Yost earned recognition for his work: the Deno House was awarded second prize in *House and Garden*'s 1945 competition for new small homes in the category of six rooms or less. This L-shaped split-level home had bedrooms located in one wing that was one-half story above the public areas, which occupied the other wing. Short flights of stairs were, according to Yost, a "livability" factor, while still providing separation of uses and privacy. The living-dining area was an open plan connected to the exterior by a porch and terrace. Most of Yost's housing designs have attached garages, designated outdoor areas connected to the public interior spaces, and basements that are only a half story below ground, allowing ample natural light to penetrate the house.

Yost also undertook demonstration projects, for clients such as the Ingersoll Steel and Disk Company, to develop a prefabricated utility unit in 1946. This unit—containing the plumbing and fixtures for the furnace, kitchen, and bath—was standard in all houses (fig. 3). Ingersoll wanted not only to test such a unit, but also to prove its flexibility for various types of architectural designs. Yost was one of eight architects commissioned to design houses for Ingersoll. His small, two-bedroom model is organized on just one floor. The prefabricated utility unit saved space and eliminated the need for a fireplace. Yet, in spite of the small size of the home, Yost managed to emphasize abundant natural light, appropriate siting for good views, an attached garage, and the terrace off the living-dining area—leit motifs of his career. This same industrialized core was used in a proposed, but unrealized, subdivision in Highland Park, Illinois, called Western Homes, which would have housed employees of the consulting firm Booz, Allen, Hamilton Company (fig. 4).[3]

In 1948 Yost was asked to design a demonstration home for the Revere Quality House Institute, a non-profit organization sponsored by the Revere Copper and Brass Company. Yost's house was the fifth in a series of prototypes built by the institute in South Bend, Indiana. In each case an architect, working in conjunction with a builder, was charged with designing and constructing a small, yet quality, house for a maximum sales price of $17,000—including the lot. Prospective homeowners could purchase

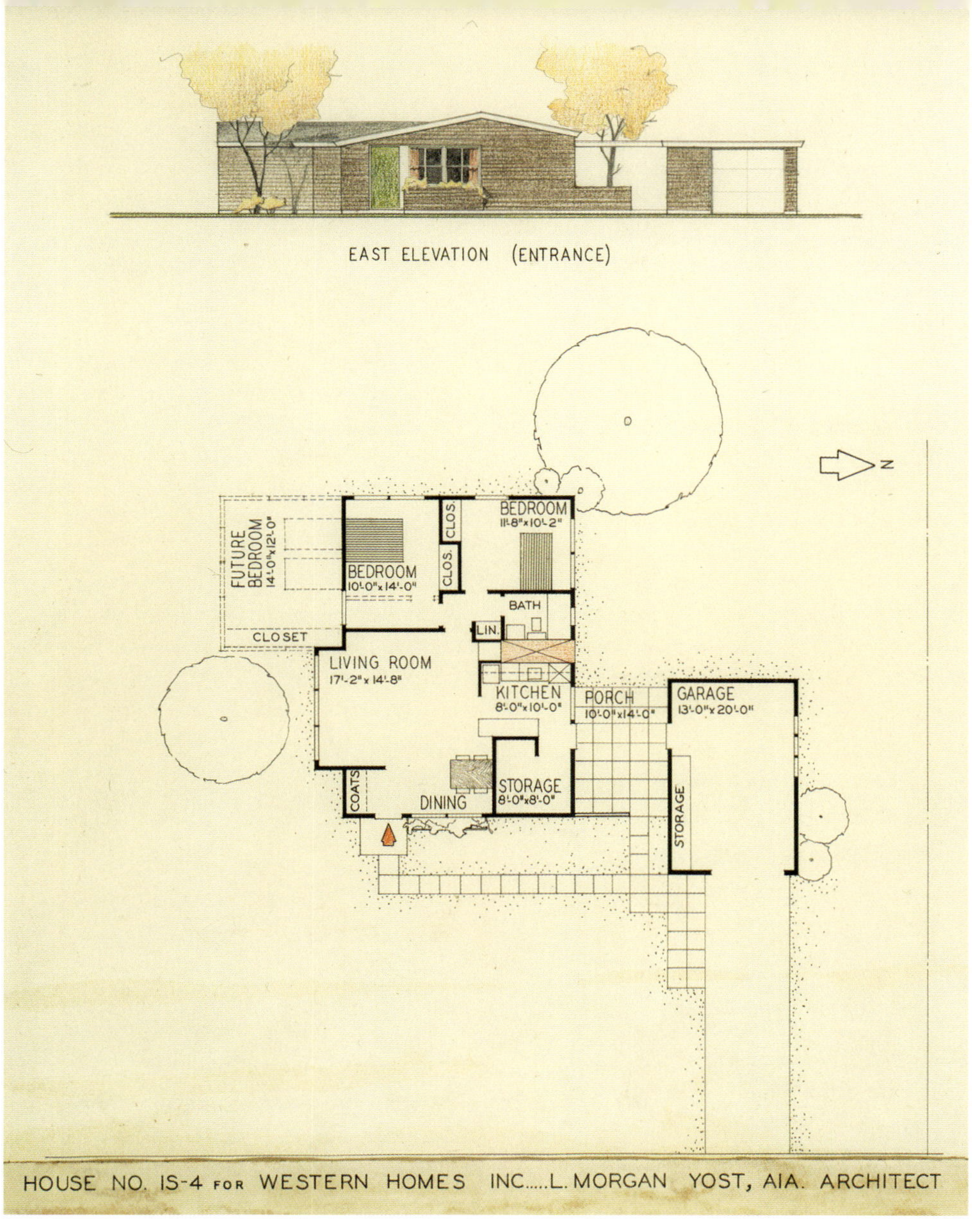

Fig. 3 (left) L. Morgan Yost. Interior perspective of a prefabricated bathroom unit from "House Studies," published in *Small Homes Guide* (Spring 1944). Ink on tracing paper; 13.5 x 13.5 cm. Gift of L. Morgan Yost, 1985.582.

Fig. 4 L. Morgan Yost. Elevation and plan of House IS-4 for Western Homes, Inc., Highland Park, Illinois, c. 1946. Delineated by Walter A. Netsch. Ink, graphite, colored pencil, and pressure-sensitive film on paper; approx. 51 x 38.2 cm. Gift of L. Morgan Yost, 1990.342.4.

a set of working drawings and specifications for the house for $100. Yost designed a thrifty rectangle approximately 77 by 20 feet with a gently sloping hip roof. The house has two basic parts, one for "polite living" and one for family play and work. Comfort is enhanced by the easy supervision of the children's activities both inside and outside from the kitchen and laundry room, large windows for plenty of light, sufficient storage, and a general ease of movement throughout the house.

Clearly, L. Morgan Yost's designs were the work of an informed and thoughtful professional. His personal touches and his consistent emphasis on quality, comfort, and rational layouts caused many of his homes to garner popularity and praise in the 1940s. However, his concerns extended beyond the work of an individual architect and reflected society's broader interests of the time. Even those ideas that were not realized during the wartime years clearly reflect the mood of a country eagerly looking toward the future. Knowing that many GIs would be returning home and understanding that the homebuilding industry could play an important part in contributing to a healthy economy, Yost made an important contribution as a writer, designer, teacher, and architect, deserving recognition as part of Chicago's architectural history and of the social history of America in the 1940s.

Notes

ZUKOWSKY 1945: Creativity on the Cusp between War and Peace, PP. 3–9

1. For Bischoff and Ertl at Auschwitz, see Robert Jan van Pelt, *The Case for Auschwitz* (Indiana University Press, 2002), pp. 295–98. For Mendelsohn at German Village, see my entry for *Mendelsohn, Erich* in John Garraty and Mark Carnes, eds., *American National Biography* (Oxford University Press, 1999), vol. 15, pp. 293–95.
2. Keith Mallory and Arvid Ottar, *The Architecture of War* (Pantheon, 1973).
3. Ellis Armstrong edited an important work titled *History of Public Works in the United States 1776–1976* (American Public Works Association, 1976) but his chapter 18 on "Military Installations" treats the subject as a general survey of places without documenting designers and planners. Examples of enthusiast books that deal with architecture during the war are too numerous to list. Some recent ones are: Graham Buchan Innes, *British Airfield Buildings of the Second World War* (Midland Publishing, 1995); Gordon Williamson, *U-Boat Bases and Bunkers 1941–45* (Osprey Publishing, 2003); Gordon L. Rottman, *Japanese Pacific Island Defenses 1941–45* (Osprey Publishing, 2003); Karl Heinz and Michael Schmeelke, *German U-Boat Bunkers Yesterday and Today* (Schiffer Military History, 1999). One of the few to mention an architect at all is Michael Foedrowitz, *The Flak Towers in Berlin, Hamburg, and Vienna 1940–1950* (Schiffer Military History, 1998). Although not concentrating on architects, one of the few books to include Russian structures is J. E. Kaufmann and Robert M. Jurga, *Fortress Europe: European Fortifications of World War II* (Combined Publishing, 1999).
4. On the design of military emblems, see Walton Rawls, *Disney Dons Dogtags* (Abbeville, 1992); on uniforms, see Paul Fussell, *Uniforms* (Houghton Mifflin, 2002); and on camouflage, see Roy M. Stanley II, *To Fool a Glass Eye* (Smithsonian Institution Press, 1998). The National Building Museum's exhibition and book on World War II illustrates camouflaged factories but does not discuss the architects or designers behind their creation: Donald Albrecht, *World War II and the American Dream* (MIT Press, 1995).
5. For a discussion of some of the German modernists who built industrial facilities during the war, see John Zukowsky, ed., *The Many Faces of Modern Architecture: Building in Germany between the World Wars* (Prestel Verlag, 1994), pp. 8, 11, 12, 17, 35–39, 239–40. For American industrial architecture of the era, see Martin Greif, *The New Industrial Landscape: The Story of the Austin Company* (Main Street Press, 1978), Grant Hildebrand, *Designing for Industry: The Architecture of Albert Kahn* (MIT Press, 1974), and *The Bridge Works: A History of Chicago Bridge and Iron Company* (Mobium Press, 1987).
6. See Brooke Kamin Rapaport and Kevin L. Stayton, *Vital Forms: American Art and Design in the Atomic Age, 1940–1960* (Brooklyn Museum of Art and Harry N. Abrams, 2002).

HANNA BOYER Bertrand Goldberg and the Legacy of 1945, PP. 10–13

1. Bertrand Goldberg, Oral History of Bertrand Goldberg, interviewed by Betty J. Blum, Chicago Architects Oral History Project, the Art Institute of Chicago, 1992, p. 118.
2. Ibid., p. 125.
3. Bertrand Goldberg, Memoirs—Presque Perdue, March 7, 1985, p. 4; an unpublished document found in the vertical files of the Department of Architecture, the Art Institute of Chicago.
4. Oral History (note 1), p. 125.
5. Bertrand Goldberg, Standard Houses Corporation transcript, p. 1; an unpublished document found in the vertical files of the Department of Architecture, the Art Institute of Chicago.
6. Oral History (note 1), p. 143.
7. Ibid., p. 143.
8. Although penicillin was discovered in 1928, it was not introduced into clinical medicine until 1941. One of the first labs to mass-produce penicillin in the United States was located in Peoria, Illinois. Because of the drug's effectiveness at fighting staph, strep, and pneumonia, it became known as the "wonder drug" of World War II. From John S. Mailer, Jr., and Barbara Mason, "Penicillin: Medicine's Wonder Drug and Its Production at Peoria, Illinois," http://www.lib.niu.edu/ipo/int810139.html.
9. Michel Ragon, *Goldberg: Dans la ville* (Paris Art Center, 1985), p. 138.
10. Oral History (note 1), p. 145.
11. Bertrand Goldberg Associates History, p. 5; an unpublished document found in the vertical files of the Department of Architecture, the Art Institute of Chicago.
12. Railway Age (Oct. 1950), p. 69; clipping from the Bertrand Goldberg scrapbook, Ryerson and Burnham Libraries, the Art Institute of Chicago.
13. Oral History (note 1), p. 153.
14. Ibid., p. 152.
15 "Plywood Boxcar Doubles as Reefer"; an unidentified clipping in the Bertrand Goldberg scrapbook, Ryerson and Burnham Libraries, the Art Institute of Chicago.
16. Oral History (note 1), p. 152.
17. "Plywood as Substitute for Metal Discussed by Defense Officials," Defense Production Record 27, no. 1 (Nov. 1, 1951).
18. "Plywood and Plastics Make Portable Home for Fast Relocation," *Wall Street Journal*, May 9, 1952.

KOWALSKI Improvisation and Inventiveness in Bruce Goff's Quonset Designs, PP. 14–17

1. Philip B. Welch, ed., *Goff on Goff: Conversations and Lecture* (University of Oklahoma Press, 1996).
2. Ibid., pp. 28–29.
3. Ibid.
4. Ibid., p. 28.
5. Peter S. Reed, "Enlisting Modernism," in *World War II and the American Dream*, ed. Donald Albrecht (National Building Museum and MIT Press, 1995).
6. "Prefabrication: Hutments to Houses," *Architectural Forum* (Feb. 1944), pp. 91–94.

7. "Navy Architecture," *Architectural Forum* (Dec. 1945), p. 74.
8. Ibid.
9. Jeffrey Cook, *The Architecture of Bruce Goff* (Harper and Row, 1978).
10. Timothy Samuelson, "Bruce Goff in Chicago," in *The Architecture of Bruce Goff, 1904–1982: Design for the Continuous Present*, ed. Pauline Saliga and Mary Woolever (Prestel and Art Institute of Chicago, 1995), p. 53.
11. Ibid.
12. "The Round House," *Life* (Mar. 19, 1951), pp. 70–76.

WOJCIK Ludwig Mies van der Rohe and Chicago in the 1940s, PP. 18–21

1. For Mies and his practice in Germany see Elaine S. Hochman, *Architects of Fortune: Mies van der Rohe and the Third Reich* (Weidenfeld and Nicolson, 1989).
2. Edward A. Duckett, *Impressions on Mies: An Interview on Mies van der Rohe: His Early Chicago Years 1938–1958*, interview with former students and associates Edward A. Duckett and Joseph Y. Fujikawa conducted by William S. Shell, 1988, p. 28.
3. See Alfred Swenson, *Pao-Chi Chang, Architectural Education at IIT 1938–1978* (Illinois Institute of Technology, 1980) and Phyllis Lambert, "Mies Immersion," in *Mies in America*, ed. Phyllis Lambert (Canadian Centre for Architecture and Whitney Museum of American Art, 2001).
4. Myron Goldsmith, Oral History of Myron Goldsmith, interviewed by Betty J. Blum, Chicago Architects Oral History Project, the Art Institute of Chicago, 1986, pp. 38–40.
5. Mies van der Rohe Archive, Part II: 1938–1967, The American Work, ed. Franz Schulze, 1992, p. 2.
6. Cited in David Spaeth, "Ludwig Mies van der Rohe: A Biographical Essay," in *Mies Reconsidered: His Career, Legacy, and Disciples*, ed. John Zukowsky (Art Institute of Chicago, 1986), p. 17.
7. Fujikawa in *Impressions on Mies* (note 2), p. 7.
8. Joseph Y. Fujikawa, Oral History of Joseph Fujikawa, interviewed by Betty J. Blum, Chicago Architects Oral History Project, the Art Institute of Chicago, 1983, p. 17.
9. Cited in Franz Schulze, "Mies van der Rohe in America," in *Chicago Architecture and Design 1923–1993: Reconfiguration of Metropolis*, ed. John Zukowsky (Art Institute of Chicago and Prestel-Verlag), p. 149.

MANNING Henry P. Glass and the Beauty of Manmade Things, PP. 22–25

1. Henry P. Glass, IDSA Living History, interviewed by Victoria Matranga, Oct. 18, 2001.
2. Kathryn B. Hiesinger and George H. Marcus, eds., *Design Since 1945* (Philadelphia Museum of Art, 1983), pp. x, xi, xii.
3. Interview with David Jameson, owner of ArchiTech gallery in Chicago, Oct. 2004.
4. Henry P. Glass, *The Shape of Manmade Things*, (E.C.G. Publishers, 1996), p. 1.
5. Ibid., p. 47.
6. Ibid., p. 159.
7. Ibid., p. 160.
8. Ibid., p. 169.
9. Transcript of speech for "Upholstery Industry," from the Henry P. Glass archive at the Art Institute of Chicago.

WOJCIK Richard Ten Eyck and Industrial Design of the 1940s, PP. 26–29

1. Arthur J. Pulos, *The American Design Adventure, 1940–1975* (MIT Press, 1988), p. 23.
2. Ibid., p. 17.
3. Ibid., p. 316.
4. John Zukowsky, "Introduction," in *Aerospace Design: Aircraft, Spacecraft, and the Art of Modern Flight*, ed. Anthony M. Springer (Merrell, 2003), p. 24.
5. Richard Ten Eyck, Oral History of Richard Ten Eyck, interviewed by Annemarie van Roessel, Chicago Architects Oral History Project, the Art Institute of Chicago, 2003, p. 29.
6. Zukowsky (note 4), pp. 21–24.
7. Phil Patton, "Man At His Best: Living Quarters, Way Cool Breeze," Richard Ten Eyck Collection, Series III: Project Papers, Box 23, Ryerson and Burnham Archives, the Art Institute of Chicago.
8. Brochure for the Aristocrat Model Sandwich Shop, Valentine Industries, Inc., Richard Ten Eyck Collection, Series III: Project Papers, Box 23, Ryerson and Burnham Archives, the Art Institute of Chicago.
9. Ten Eyck (note 5), p. 40.

THORNE L. Morgan Yost's Houses for Modern Living, PP. 30–33

1. L. Morgan Yost. "This Business of Being Busy," *The Archi* 23, no. 1 (Nov. 1941), pp. 1–3ff.
2. The exhibition, sponsored by the American Institute of Architects, was held at the Art Institute during January and February 1942 and displayed Chicago-area designs selected by a jury.
3. The subdivision was planned for two- and three-bedroom homes. Although all houses were to contain a standard plumbing core, nine different models were designed to ensure variety.

Acknowledgments

The idea for this project came from a 2001 discussion with Martha Thorne and Clive Dilnot in which we concluded that it would be appropriate for the Art Institute's Department of Architecture to commemorate the sixtieth anniversary of the end of World War II with an exhibition that focused on architecture and design of the forgotten forties, a decade that has been eclipsed by books and exhibitions on Art Deco modernism of the 1930s and International Style modernism of the 1950s. The first planning sessions for this project began in 2002, and they included local academics from the University of Chicago, the Illinois Institute of Technology, the Institute of Design, the School of the Art Institute of Chicago, and the University of Illinois at Chicago; museum personnel from the First Division Museum at Cantigny in Wheaton, the Chicago Historical Society, and the Pritzker Military Library; representatives from the armed forces, the City of Chicago, veterans' groups such as the Chicago Navy League, the Illinois LST Association, the DODO Chapter of the Tuskegee Airmen, and the First Fighter Association; and special interest groups such as the Chicago Maritime Heritage Society, the Experimental Aircraft Association's Warbirds Squadron 4, and the Windy City Chapter of Women in Aviation. All actively participated in the creation of this project. Of those groups, I especially want to thank Jim Brewer and Capt. John Redfield of the Air Force Public Affairs Office in Chicago, Mark Laatsch of the Warbirds, Kim Sailor of Women in Aviation, and Dave Truitt of Chicago Maritime Heritage for their consistent support of our endeavors. The project also had the active daily attention of the staff of the Art Institute's Department of Architecture, and its dedicated volunteers such as Barbara Ahlberg, Phil Kennedy, Allison Klein, Silke Oldenburg, Sonja Ostendorf, and Marta Wojcik. Marta and Carissa Kowalski of the Department of Architecture in particular deserve my special thanks for taking this project to successful completion in 2005. In the Publications Department I would like to thank Sarah Guernsey, Robert Sharp, and Elizabeth Stepina. Jeff Wonderland created the book's design. But, of all the people who assisted us in this endeavor, I want to thank veterans such as Bob Bridgeman of the LST Association and Bev Dunjill of the Tuskegee Airmen, wartime detainees such as Eberhard Fuhr, and my own parents, for sharing their ideas and experiences of this era, and inspiring me to pursue this project.

JOHN ZUKOWSKY

Contributors

JOHN ZUKOWSKY is Director of the Westcott House Foundation in Springfield, Ohio, and was formerly the John H. Bryan Curator of Architecture at the Art Institute of Chicago. His recent publications and exhibitions at the Art Institute cover a wide range of subjects, including aerospace design and Chicago architecture.

LORI HANNA BOYER is the Collection Manager for the Department of Architecture at the Art Institute of Chicago, and she was formerly the Bertrand Goldberg Project Director for the department. She holds an M.A. in Art History from the University of Missouri–Columbia.

CARISSA KOWALSKI works in the Department of Architecture and is a master's degree candidate in Art History at the University of Illinois at Chicago. She has also been involved with writing, researching, and editing several publications on such diverse topics as Chicago architecture, psychological assessments, and jazz album cover design.

MARTA WOJCIK is a former research intern in the Department of Architecture who worked closely with John Zukowsky on this exhibition and its related programs. She received an M.A. in Art History from Jagiellonian University in Krakow, Poland.

KAY MANNING is a veteran editor who works for a Chicago daily newspaper. She recently received a graduate certificate in museology and has been assisting the Department of Architecture at the Art Institute of Chicago for the past two years.

MARTHA THORNE is Associate Curator and Acting Head of the Department of Architecture at the Art Institute of Chicago. She received a Master of City Planning degree from the University of Pennsylvania. She joined the Art Institute in 1996 and has undertaken many exhibitions and publications related to contemporary architecture.